How to Know the Truth Absolutely

by Rolf A. F. Witzsche

Contents

On the Ice Age and Climate Change
and the book

How to Know the Truth Absolutely

To know the truth absolutely! That's often deemed impossible. However, the notion is false. What is truth, is knowable.

The book presented here illustrates how it is possible for a person to solve complex problems, such as to double a square, in area, and than to know in the end that the result is absolutely correct; and all this without one having any special knowledge in mathematics and geometry. Plato illustrated how this is done in his 'Meno Dialog' in the 'Republic.'

Now extend the process further and build on the 'Meno' principle to prove the Pythagorean theorem. At first glance, this seems impossible. However, when one extends the 'Meno' principle, by understanding its expression, one can say with absolute certainty that the result is correct by simply looking at the process with which the answer is derived.

Plasma in the physical universe is as challenging in perception as the spiritual domain in the human sphere. Both are invisible, except by their effects, but they are understandable and knowable.

On the same basis it becomes possible to recognize the outcome of plasma physics operating in the universe. This is important in astrophysics, because of the huge consequences that follow, when we mess up in our thinking.

With the Ice Age Challenge now before us, we face two imperatives. One is to understand the physical dynamics, and to create the physical infrastructures that enable human living to continue in an Ice Age climate. The second challenge, the greater challenge, is to raise up our humanity to such height as will impel us to get the job done. Some say that miracles are needed on both fronts. But what of it? Are we, as human beings, not the miracle makers on the Earth?

In the real universe, the cosmic operations are anti-entropic in nature, and expanding and progressing. We, ourselves are evidence of this progression. Should this progression have ended? Neither is our Sun isolated from the progressive nature of the universe, but expresses its dynamics, its resonating plasma streams, and their reflection in the climate on Earth. Shouldn't we develop ourselves spiritually and culturally, likewise?

Climate Change reflects the nature of the universe. It should also be reflected in us.

The Earth itself is the creation of the Sun, with its atoms having been massively synthesized in high-energy times near the center of the galaxy.

The synthesizing plasma fusion is presently at a low state, though it is currently enhanced for our Sun by electromagnetic 'Primer Fields' that focus interstellar plasma onto the Sun in a highly condensed manner. When the plasma-focusing system becomes inactive, below the required threshold conditions, the Sun reverts to a type of cosmic default level with 70% less energy being radiated, and higher rates of solar cosmic-ray flux being experienced.

At the present rate of plasma diminishment being experienced, the solar activity phase-shift threshold to the next Ice Age period may be crossed in 30 years, or in the 2050s, most likely. With the primer-fields system gone inactive by then, the climate on Earth will get 40 times colder than the Little Ice Age in the 1600s had been. Ice core evidence promises that. Without the needed preparations for human living in such an environment, 99% of humanity would die of starvation, both by the cold, and by CO_2 depletion that diminishes agriculture, as more CO_2 becomes dissolved into the sea.

With the 'Primer Fields' being critical for our very existence, the exploration of them is likewise critical.

In the Little Ice Age, between 10% and up to 30% of the populations in Europe had perished by starvation. The last Big Ice Age was evidently vastly harsher. Only 1-10 million people emerged from it alive. That's all we had after 2 million years of development. We want to do far better this time around; and we can, with large-scale technological infrastructures for our food supply. But will we create them? Will we get the job done in the 30 years that we still have left before the Ice Age starts anew? Will we even consider it? And how certain are we that the phase shift to the next glaciation period will begin, as the evidence suggests, in the 2050s? We have no slack on this front. Should we fail us on this absolute front, we would be committing suicide.

Numerous fields of evidence tell us that the next Ice Age is near. That's where the truth begins. Most of the evidence was discovered in the 1990s and thereafter. Some evidence is measured in ice cores; some is measured in space, by satellites. Some measurements are also made on the ground in terms of measurements of the Earth's magnetic-pole drift observed in northern Canada. All of this is seen combined with high-energy physics experiments at a leading national laboratory, and is also explored in the small in static experiments.

So, what will the answer be? Will we move with the evidence? Or will we lay ourselves down to die by default?

It takes an independent researcher to brake the taboos that have kept mainstream cosmology imprisoned, increasingly, during the past century, even while what is regarded as taboo is known to be wrong.

The Illustrated Science series is intended to open the scene beyond the threshold of accepted taboos, to where the actual physical evidence speaks for itself.

The scope of the existential challenge that the Ice Age brings with it, takes astrophysics out of the academic domain and places it into the foreground as one of the most-critical issues of our time. The big Climate Change events that have already worldwide effects are mere fringe effects in the flow of the ever-changing cosmic dynamics. The big effect, when the Ice Age begins anew, promises to be caused by a dimmer and colder Sun. The loss of 70% of the Sun's radiated energy defines our climate future that begins in the near term.

Sure, we can live with all that by creating new platforms for agriculture that are able to operate under Ice Age conditions. But will we do it? The task is enormous. Or will we fail ourselves on this front? We have no reason to allow us to fail. We have the materials and energy resources on hand to accomplish everything that is required for us to continue to live in an Ice Age World. But will we do it? The big question that never goes away, therefore, is; will we develop our inner resources as human beings sufficiently to get the job done, and to get it done in time? Or will we do nothing, ignore the challenge, and condemn our children and one-another to an agonizing death by starvation? That's the choice.

Towards meeting the inner challenge, I have created the epic series of novels, The Lodging for the Rose. And further, towards meeting the science challenge, I have produced numerous research books and several dozen exploration videos that the Illustrated Science series is modeled after. The work is the result of a quarter century of research, for which numerous elements of evidence in related fields came to light during the timeframe of my research.

It is my hope that the work that went into all of these projects will help in some degree - for humanity that we are all a part of - to write itself a ticket to have a future.

High-resolution color images, of the images in this book, can be obtained at www.iceagetheatre.ca

The Pythagorean theorem

Pythagoras of Samos

Bust of Pythagoras in the Capitoline Museums, Rome - wikipedia

In the late 6th Century BC, the great Ionian Greek philosopher, mathematician, and founder of a religious society, made an amazing discovery in geometry that bears his name to the present day, termed the Pythagorean theorem.

Most of the information about him was written down centuries after he lived, so that many of the accomplishments he is credited with may actually have been accomplishments of his colleagues and successors. Nevertheless, the Pythagorean ideas had a significant influence on Plato, and through him, on all of Western philosophy.

It is said that Pythagoras was the first man to call himself a philosopher, a lover of wisdom. This means that his famous work, the Pythagorean theorem, is more than just a discovery in geometry, but involves a form of science that enables one to understand aspects of truth that are hidden from the senses, but can be recognized in the mind to be absolutely truthful, and this without the aid of mathematics.

A theorem in geometry

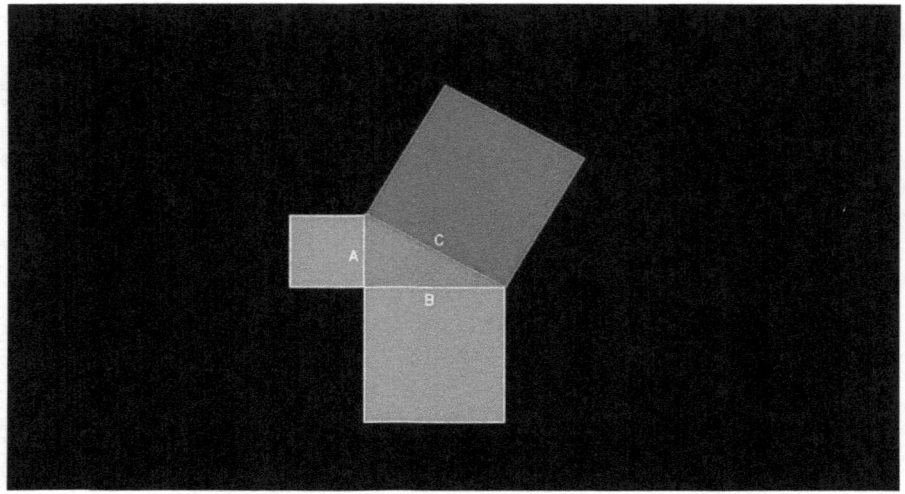

So, what is the Pythagorean theorem all about? Simply put, it is a theorem in geometry that states that in a right-angled triangle the area of the square on the longest side, the side opposite to the right angle, is equal to the sum of the areas of the squares of the other two sides.

At first glance, no rational reason is apparent that this should be the case. Of course we all know that the theorem is correct. It is taught in all the schools, isn't it? But how do we know this to be so, with absolute certainty? Pythagoras might have answered that it is possible for us to know this, because we are human beings that have the capability to see with the mind. When we do this, we also discover a bit more of what a human being is. And so, Pythagoras would say, "prove it to yourself that I am right," and thereby discover who you are as a human being.

Socrates begins one of the most influential dialogues

On this basis Socrates begins one of the most influential dialogues of Western philosophy, regarding the argument for the inborn capacity in humanity, to know truth, which he calls virtue. He argues that there is no such thing as teaching and learning, but only remembering.

Plato may have been a 'student' of Socrates

Plato in his academy, drawing after a painting by Swedish painter Carl Johan Wahlbom

Plato may have been a 'student' of Socrates, being almost 40 years younger. Plato conveyed some of the concepts of Socrates in his writings.

Among Plato's writing is the famous Meno dialogue

Plato in his academy, drawing after a painting by Swedish painter Carl Johan Wahlbom

Among Plato's writing is the famous Meno dialogue. In the dialog, in a similar academy setting, Socrates tells his friend Menon that even a boy who has no training in geometry and mathematics, is able to understand complex problems, and is able to understand with absolute certainty that what he understands is correct.
Socrates asks his friend Menon to select someone from his workers, and that he would prove his case in him.

Socrates drew a 2-foot wide square on the ground

To proof his case, Socrates drew a 2-foot wide square on the ground, and proposed to the boy, whom Menon has selected, that he is able to create a larger square from it, that covers double the area.

He proposed to the boy

He proposed to the boy that if he drew another square of equal size beside it, the result would cover double the area, but it wouldn't be a square anymore. So how does one solve the puzzle?

Socrates proposed that if one adds

Socrates proposed that if one adds half the original square to either side of it, the result covers double the area, as required, and that its shape more closely resembles a square, but still falls short of being a square. Consequently, this approach doesn't work either, does it?

Then Socrates proposed to the boy

Then Socrates proposed to the boy that he should draw a large square that is 4 times as big as the original square.

Then he asks that he divide each of the four squares in half

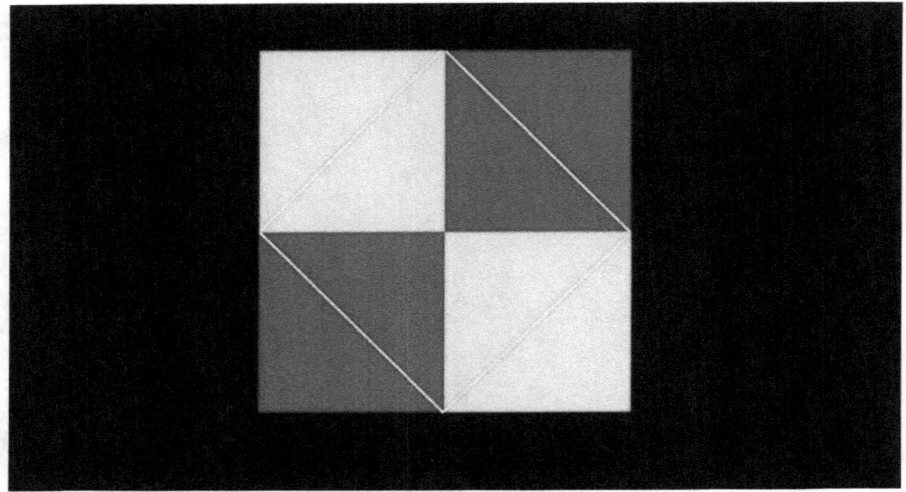

Then he asks the boy that he divide each of the four squares in half, by drawing a line from corner to corner, and that he does this in such a manner that the dividing lines together, form a square.
Once it was don, he asked the boy in essence: Is the resulting inner square, not twice as large in area as the original square?
The boy agrees that the inner square is twice as large, because the inner square contains 4 triangles, while the original square contains only 2 triangles.
Socrates asked then, "Can you say then with absolute certainty that the inner square is twice as big in area, than the original square?"
Of course it is, the boy answers. "The original square contains 2 triangles, the new square contains 4. It is twice as big in area. It's as simple as that."

In principle a proof of the Pythagorean theorem

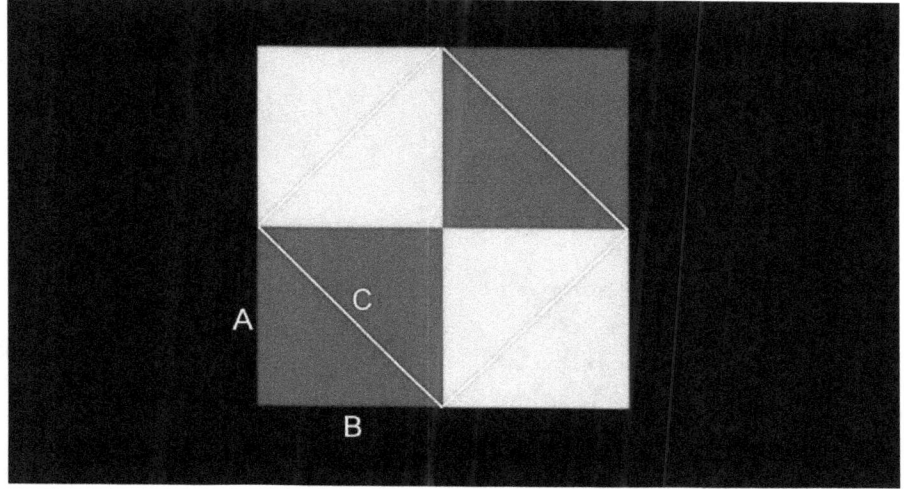

Socrates might have added, speaking to his friend Menon, that the boy has just delivered in principle a proof of the Pythagorean theorem. The theorem states that the squares over the small sides of a right triangle, when added together, are equal in area to the square over the long side. In the case at hand, A-square is the size of the original square, likewise B-square. The two added together, add up to C-square that is twice as big. The boy proved Pythagorean theorem. Of course, the proof, in this particular case is simple, as A and B are of equal length.

The principle applies to all cases of a right triangle

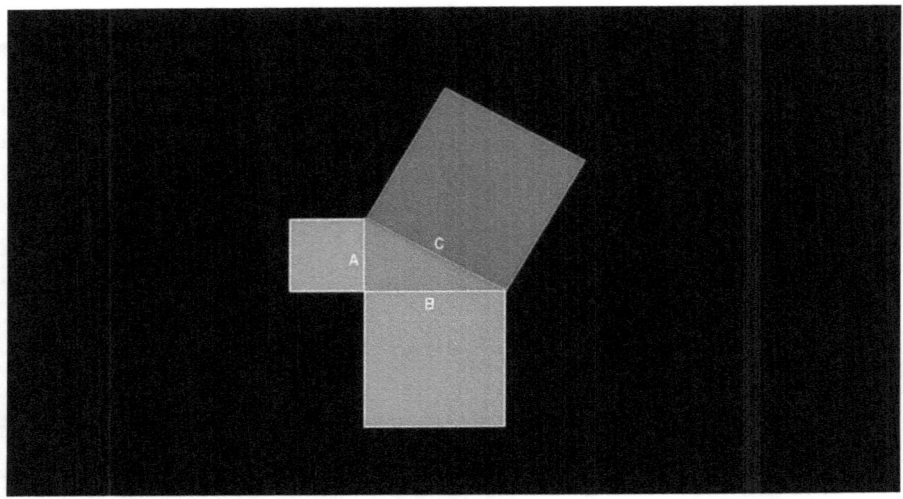

Socrates might ask his friend further whether the principle that is illustrated applies to all cases of a right triangle, no matter what their shape may be.

Can we answer this question for him?

I think that Pythagoras would expect this of us. But how can we prove this to be so?

By simply extending the principle presented in the Meno dialog

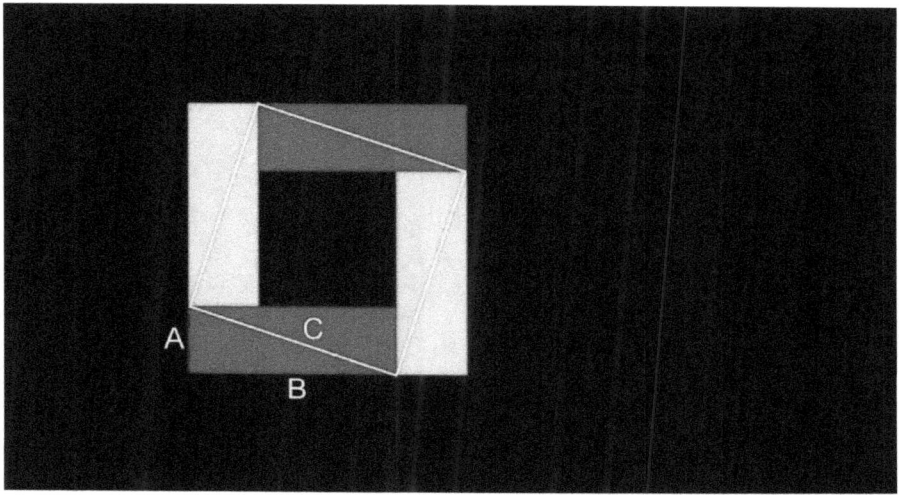

It is possible to do this fairly easily, by simply extending the principle presented in the Meno dialog, and applying it to all right triangles. In this case one would mirror the triangles into rectangles, and put four of them together, as in the Meno dialog. One would thereby simply stretch the squares of the Meno dialog, into rectangles. Would this be enough to prove the point?
Pythagoras proposed that the area of C-square is equal in size to the areas of A-square and B-square added together. Can we prove this to be so in this extended case?
Certainly we can. We can proof this the same way as before.

We can do this by counting triangles

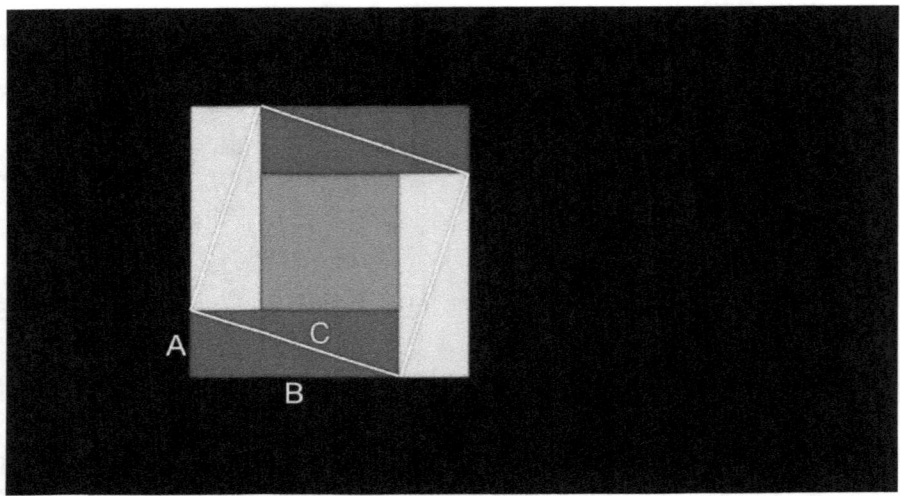

We can do this by counting triangles. The C-square is again made up of 4 triangles, plus the big red area in the middle. I propose that we now create a construct that is twice as large in area as C-square, and proof to ourselves that we can place two A-squares, and two B-squares into it. If we can do this, we have delivered proof of the Pythagorean theorem for all cases.

The proof lies in that we can prove that C-square covers precisely half of the area that we can place two A-squares, and two B-squares into.

So, let's see if we can do this

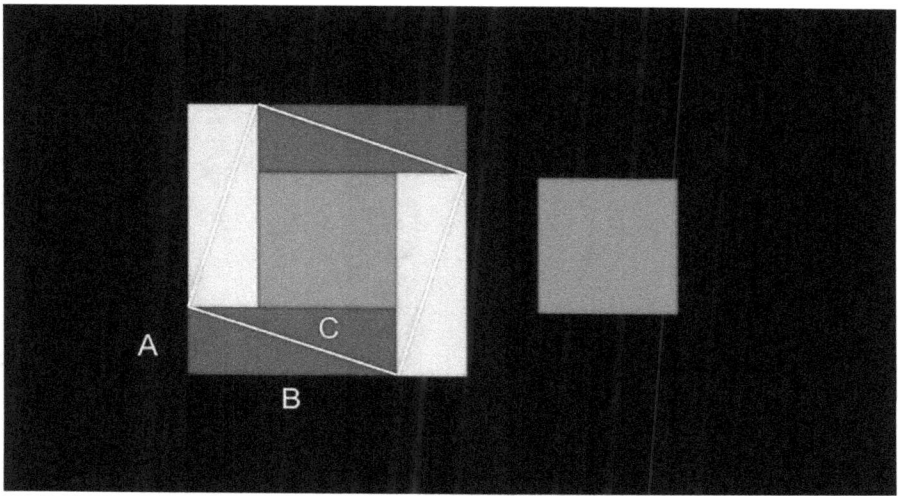

So, let's see if we can do this. Since C-square is made up of 4 triangles and the red center, we will have to create an area that contains 8 triangles, and in addition another square that is equal to the red center. The result is shown here.

23

Now let us see if this works

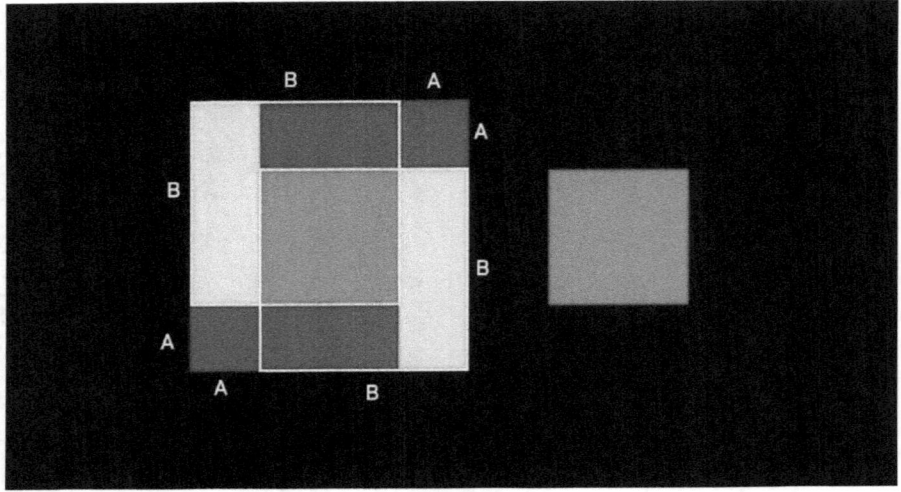

Now let us see if this works. Can we place two A-squares and two B-squares into this larger area. That's not hard to do, is it? We can place two of the big B-squares into the corners, drawn in yellow. As you can see, the two squares overlap the red center. The overlapped area happens to be equal in size to the red square set on the side. With this done, the remaining areas are of the exact size that we need to fit two A-squares into. Thus, the requirement that we set out to achieve, has been achieved with absolute precision.

By following the simple process that Socrates has started, we have discovered a principle of proof that applies to every possible shape of right triangles. And as I promised, no mathematical knowledge was needed to prove to ourselves that what Pythagoras had discovered more than 2,500 years ago is absolutely correct in all possible cases.

Few people know why the principle is correct

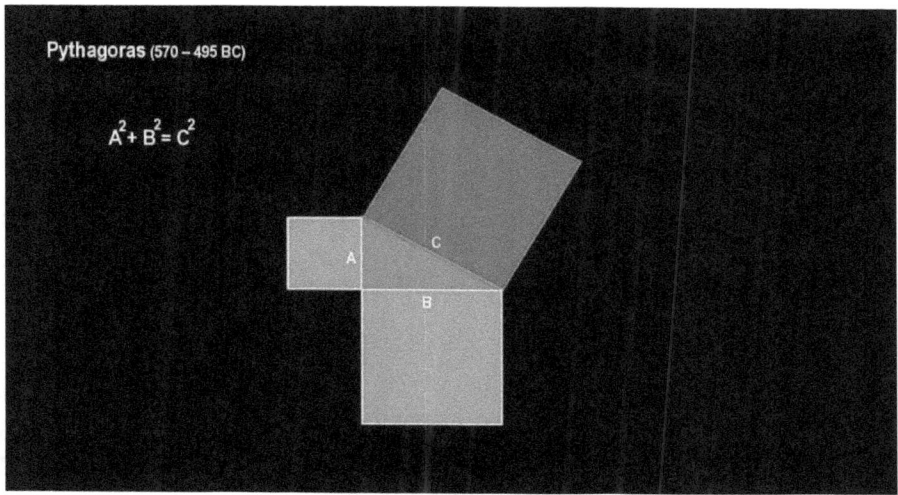

Pythagoras (570 – 495 BC)

$$A^2 + B^2 = C^2$$

In the schools we are taught the mathematical formula only, that A-square plus B-square equals C-square. The formula is useful, of course, for practical application. Indeed, the formula is routinely applied on a wide horizon of applications. But few people know why the principle that the theory reflects, is correct. They accept the formula on faith. That's what gets society into dangerous waters.

Faith without understanding has become a huge trap

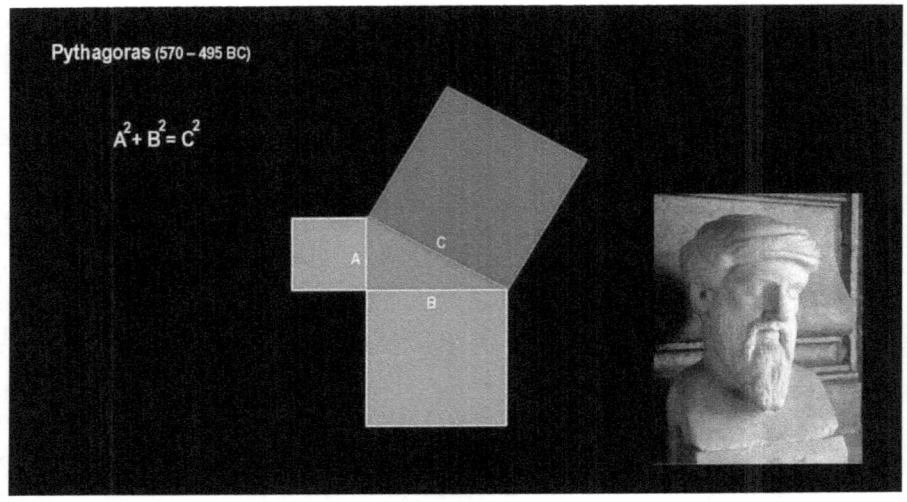

Pythagoras (570 – 495 BC)

$$A^2 + B^2 = C^2$$

We don't know by what reasoning Pythagoras had developed his theorem, because everything we know about him was written down centuries after he had lived. From his time to ours, several hundred different ways have been developed to prove his theorem. Most of the methods to prove the theorem are extremely complicated, which they don't need to be. As far as I know, not a single one of the published methods is as simple as the one I have presented here, built on the Meno principle that enables a person with no special knowledge in mathematics and geometry, to simply look at the problem with the eye of the mind, and thereby to discover with absolute certainty, based on a few obvious facts, that Pythagoras was right. We are able to discover this amazing proof, because the quality of reason that it results from, is built into our humanity In the proof that I have presented here, not a single aspect of it needs to be taken on faith, whether it be faith in geometry or faith in mathematics. And this is where its value lies, because when faith takes the place of understanding, humanity is rushing into a trap. Faith without understanding has become a huge trap in modern time, on many fronts, including in critical aspects of science.

Much of the world now lives on faith

Much of the world now lives on faith. People's faith is often carefully guided for political objectives, typically to hide the truth.

In Lies We Trust

"In Lies We Trust" has become a widely celebrated sport, a sport of forcing war..

Sometimes the doctrines are openly brutal

Sometimes the doctrines are openly brutal, that demand society's acceptance on faith. The outcome is rarely less than exceedingly tragic.

The global warming doctrine

The global warming doctrine for which humanity is now burning its own food in the midst of mass-starvation, is another tragic example.

All plainly false theories

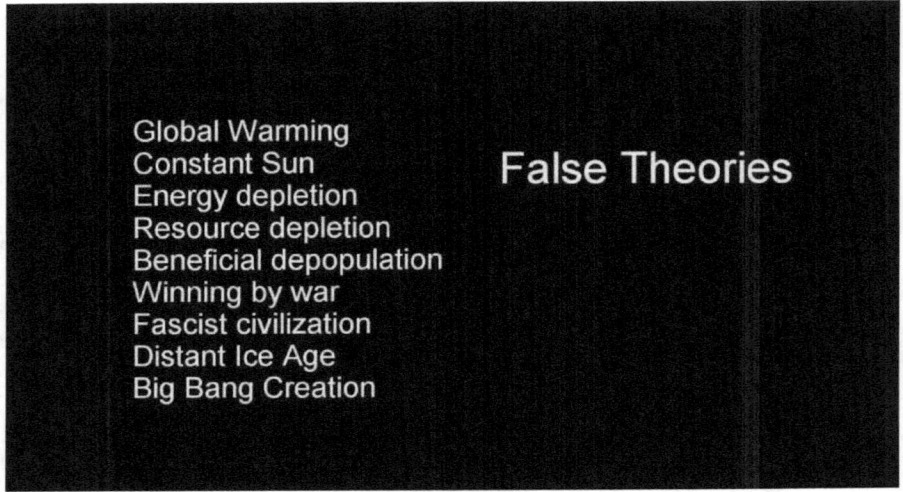

Similarly, tragic is the doctrine of the constant Sun, and of the far distant Ice Age, and the entropic universe doctrine of the Big Bang theory, and so on, and on, which are all plainly false theories that simple physical evidence proves to be false in every case. This does not mean that the truth cannot be recognized with absolute certainty, when the principles for the truth are explored and understood.

False doctrines have all one factor in common

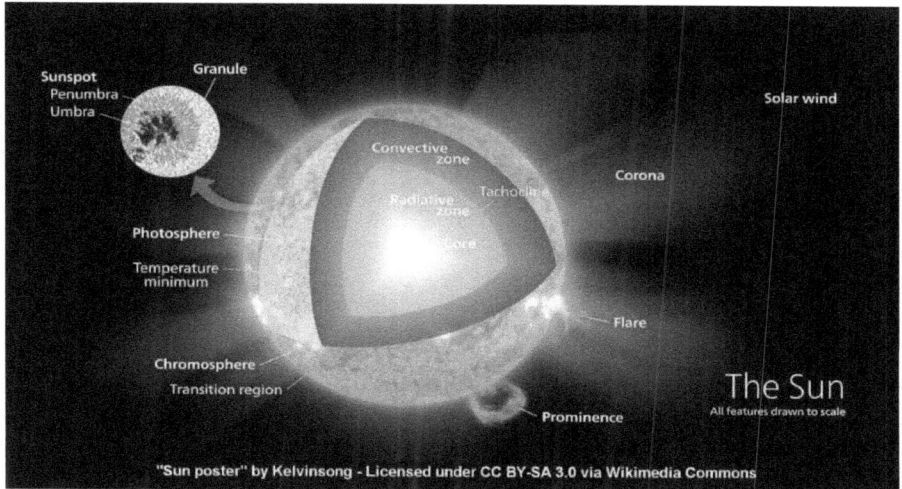

"Sun poster" by Kelvinsong - Licensed under CC BY-SA 3.0 via Wikimedia Commons

As one might expect, the false doctrines have all one factor in common, they require faith in mysteries for which no evidence exists, while contrary evidence does exist that adds up to monumental items of proof.

Plato gained his fame for his searching for the truth

Plato gained his fame for his unyielding searching for the truth, raising humanity up with profound discoveries to great freedoms, shown here in the fresco of the Academy of Athens by Rafael.

Plato is pointing upwards

Plato is pointing upwards. Aristotle, in contrast, puts humanity down, and keeps it small by saying that there exists no such thing as truth, that all is mere opinion.

Under the banner 'In Lies We Trust,' faith in lies has unleashed a series of bitter dark ages in which civilization was destroyed and society dehumanized and brutalized.

Nevertheless, the truth does exist. We find it far and wide supported with monumental evidence that can be understood with the same absolute certainty with which the slave boy in the Meno dialog proved in principle the Pythagorean theorem. Once we break away from Aristotle's trap of faith without reason and understanding, to Plato's freedom in the truth based on reason and understanding, we are on the way to the greatest renaissance of all times. This renaissance in our age would unfold as the greatest Strategic Defense Initiative (SDI) for the defence of humanity that is urgently needed, which would enable us to prepare our world for the next Ice Age that is on the near horizon in potentially the 2050s timeframe.

The resulting strategic defence of humanity based on truth built on evidence and discovered principles, would with great certainty inspire a new sense of humanity in society that would eradicate empire and its poverty, its depopulation policy and nuclear war, together with its destructive monetarism along the way, as these would fall away as lesser challenges.

All of this is immanently possible

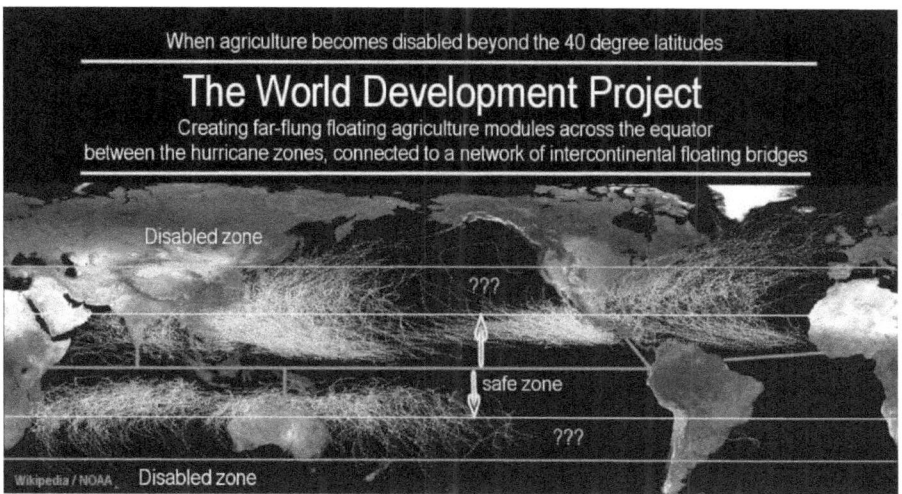

And all of this is immanently possible if we care to make it so, because humanity is inclined in its very soul to give itself this chance for the simple reason that the process of discovering and of valuing the truth, is a built-in factor of our humanity that cannot forever remain denied.

www.ingramcontent.com/pod-product-compliance
Lightning Source LLC
Chambersburg PA
CBHW070423190526
45169CB00003B/1385